OFFICIAL SQA PAST PAPERS WITH ANSWERS

D0255730

STANDARD GRADE | CREDIT

# MATHEMATICS
## 2006-2010

© Scottish Qualifications Authority

First exam published in 2006.
Published by Bright Red Publishing Ltd, 6 Stafford Street, Edinburgh EH3 7AU
tel: 0131 220 5804 fax: 0131 220 6710 info@brightredpublishing.co.uk  www.brightredpublishing.co.uk

ISBN 978-1-84948-100-7

A CIP Catalogue record for this book is available from the British Library.

Bright Red Publishing is grateful to the copyright holders, as credited on the final page of the book, for permission to use their material.
Every effort has been made to trace the copyright holders and to obtain their permission for the use of copyright material.
Bright Red Publishing will be happy to receive information allowing us to rectify any error or omission in future editions.

## STANDARD GRADE | CREDIT

# 2006

[BLANK PAGE]

C

# 2500/405

NATIONAL
QUALIFICATIONS
2006

FRIDAY, 5 MAY
1.30 PM – 2.25 PM

MATHEMATICS
STANDARD GRADE
Credit Level
Paper 1
(Non-calculator)

1 **You may NOT use a calculator**.

2 Answer as many questions as you can.

3 Full credit will be given only where the solution contains appropriate working.

4 Square-ruled paper is provided.

**FORMULAE LIST**

The roots of $ax^2 + bx + c = 0$ are $x = \dfrac{-b \pm \sqrt{(b^2 - 4ac)}}{2a}$

**Sine rule:** $\dfrac{a}{\sin A} = \dfrac{b}{\sin B} = \dfrac{c}{\sin C}$

**Cosine rule:** $a^2 = b^2 + c^2 - 2bc \cos A$ or $\cos A = \dfrac{b^2 + c^2 - a^2}{2bc}$

**Area of a triangle:** Area $= \frac{1}{2} ab \sin C$

**Standard deviation:** $s = \sqrt{\dfrac{\sum (x - \bar{x})^2}{n-1}} = \sqrt{\dfrac{\sum x^2 - (\sum x)^2 / n}{n-1}}$, where $n$ is the sample size.

KU | RE

**1.** Evaluate

$$56\cdot4 - 1\cdot25 \times 40.$$

2

**2.** Evaluate

$$1\tfrac{3}{5} + 2\tfrac{4}{7}.$$

2

**3.** Given that $f(x) = 4 - x^2$, evaluate $f(-3)$.

2

**4.**

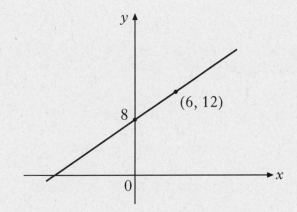

Find the equation of the given straight line.

3

**[Turn over**

**5.** (*a*) Factorise

$$4x^2 - y^2.$$

1

(*b*) Hence simplify

$$\frac{4x^2 - y^2}{6x + 3y}.$$

2

**6.** Solve the equation

$$x - 2(x + 1) = 8.$$

3

**7.** Coffee is sold in regular cups and large cups.

The two cups are mathematically similar in shape.

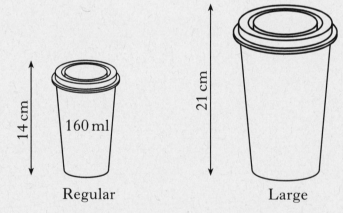

The regular cup is 14 centimetres high and holds 160 millilitres.

The large cup is 21 centimetres high.

Calculate how many millilitres the large cup holds.

4

**8.** The graph of $y = x^2$ has been moved to the position shown in Figure 1.

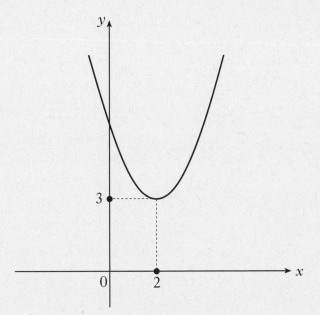

Figure 1

The equation of this graph is $y = (x-2)^2 + 3$.

The graph of $y = x^2$ has now been moved to the position shown in Figure 2.

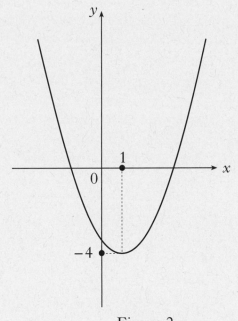

Figure 2

Write down the equation of the graph in Figure 2.

**[Turn over**

9. Euan plays in a snooker tournament which consists of 20 games.

He wins $x$ games and loses $y$ games.

(a) Write down an equation in $x$ and $y$ to illustrate this information.

1

(b) He is paid £5 for each game he wins and £2 for each game he loses.

He is paid a **total** of £79.

Write down another equation in $x$ and $y$ to illustrate this information.

2

(c) How many games did Euan **win**?

3

10. Triangle ABC is right-angled at B.

The dimensions are as shown.

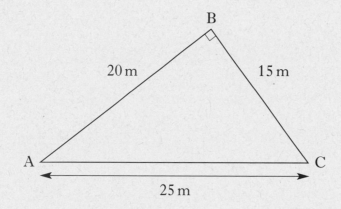

(a) Calculate the area of triangle ABC.

1

(b) BD, the height of triangle ABC, is drawn as shown.

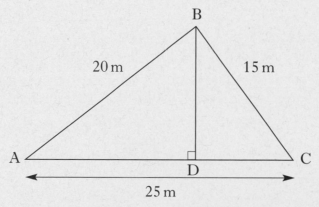

Use your answer to part (a) to calculate the height BD.

3

**11.** (a) One session at the Leisure Centre costs £3.

£3 per session

Write down an algebraic expression for the cost of $x$ sessions.

(b) The Leisure Centre also offers a monthly card costing £20. The **first 6** sessions are then free, with each additional session costing £2.

Monthly card
£20

* first 6 sessions free
* each additional
  session £2

(i) Find the **total** cost of a monthly card and 15 sessions.

(ii) Write down an algebraic expression for the **total** cost of a monthly card and $x$ **sessions**, where $x$ is greater than 6.

(c) Find the minimum number of sessions required for the monthly card to be the cheaper option.

**Show all working.**

*[END OF QUESTION PAPER]*

[BLANK PAGE]

**C**

# 2500/406

NATIONAL
QUALIFICATIONS
2006

FRIDAY, 5 MAY
2.45 PM – 4.05 PM

## MATHEMATICS
## STANDARD GRADE
Credit Level
Paper 2

1 **You may use a calculator**.

2 Answer as many questions as you can.

3 Full credit will be given only where the solution contains appropriate working.

4 Square-ruled paper is provided.

SCOTTISH
QUALIFICATIONS
AUTHORITY

©

## FORMULAE LIST

The roots of $ax^2 + bx + c = 0$ are $x = \dfrac{-b \pm \sqrt{(b^2 - 4ac)}}{2a}$

**Sine rule:** $\dfrac{a}{\sin A} = \dfrac{b}{\sin B} = \dfrac{c}{\sin C}$

**Cosine rule:** $a^2 = b^2 + c^2 - 2bc \cos A$ or $\cos A = \dfrac{b^2 + c^2 - a^2}{2bc}$

**Area of a triangle:**     Area $= \frac{1}{2}ab \sin C$

**Standard deviation:** $s = \sqrt{\dfrac{\sum(x - \bar{x})^2}{n - 1}} = \sqrt{\dfrac{\sum x^2 - (\sum x)^2 / n}{n - 1}}$, where $n$ is the sample size.

KU | RE

1. The orbit of a planet around a star is circular.

   The radius of the orbit is $4 \cdot 96 \times 10^7$ kilometres.

   Calculate the circumference of the orbit.

   Give your answer **in scientific notation**.

   3

2. (a) The pulse rates, in beats per minute, of 6 adults in a hospital waiting area are:

   68    73    86    72    82    78.

   Calculate the mean and standard deviation of this data.

   3

   (b) 6 children in the same waiting area have a mean pulse rate of 89·6 beats per minute and a standard deviation of 5·4.

   Make **two** valid comparisons between the children's pulse rates and those of the adults.

   2

3. Harry bids successfully for a painting at an auction.

   An "auction tax" of 8% is added to his bid price.

   He pays £324 in total.

   Calculate his bid price.

   3

   **[Turn over**

**4.** (*a*) Expand and simplify

$$(x+4)(3x-1).$$

(*b*) Expand

$$m^{\frac{1}{2}}(2+m^2).$$

(*c*) Simplify, leaving your answer as a surd

$$2\sqrt{20}-3\sqrt{5}.$$

**5.** ST, a vertical pole 2 metres high, is situated at the corner of a rectangular garden, PQRS.

RS is 8 metres long and QR is 12 metres long.

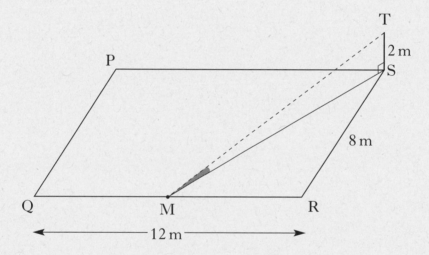

The pole casts a shadow over the garden.

The shadow reaches M, the midpoint of QR.

Calculate the size of the shaded angle TMS.

KU | RE

**6.** (*a*) There are three mooring points A, B and C on Lake Sorling.

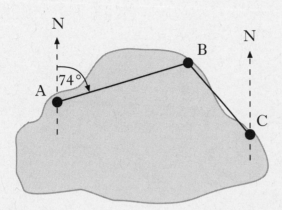

From A, the bearing of B is 074°.

From C, the bearing of B is 310°.

Calculate the size of angle ABC.

2

(*b*) B is 230 metres from A and 110 metres from C.

Calculate the direct distance from A to C.

Give your answer **to 3 significant figures**.

4

**7.** (*a*) A block of copper 18 centimetres long is prism shaped as shown.

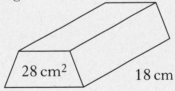

The area of its cross section is 28 square centimetres.

Find the volume of the block.

1

(*b*) The block is melted down to make a cylindrical cable of diameter 14 **millimetres**.

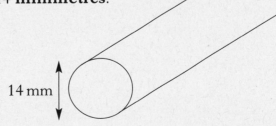

Calculate the length of the cable.

4

KU | RE

8. A set of scales has a circular dial.

The pointer is 9 centimetres long.

The tip of the pointer moves through an arc of 2 centimetres for each 100 grams of weight on the scales.

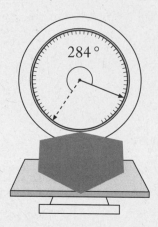

A parcel, placed on the scales, moves the pointer through an angle of 284°.

Calculate the weight of the parcel.

4

9. The number of diagonals, $d$, in a polygon of $n$ sides is given by the formula

$$d = \tfrac{1}{2} n (n-3).$$

(a) How many diagonals does a polygon of 7 sides have?

2

(b) A polygon has 65 diagonals.

Show that for this polygon, $n^2 - 3n - 130 = 0$.

2

(c) Hence find the number of sides in this polygon.

3

KU RE

**10.** Emma goes on the "Big Eye".

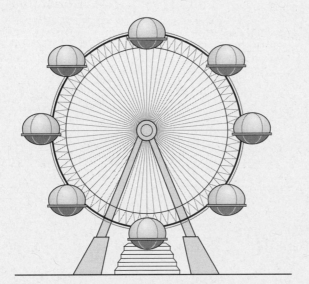

Her height, $h$ metres, above the ground is given by the formula

$$h = -31\cos t° + 33$$

where $t$ is the number of seconds after the start.

(a) Calculate Emma's height above the ground 20 seconds after the start.    2

(b) When will Emma first reach a height of 60 metres above the ground?    3

(c) When will she next be at a height of 60 metres above the ground?    1

**[Turn over for Question 11 on *Page eight***

**11.** In triangle ABC,

BC = 8 centimetres,

AC = 6 centimetres and

PQ is parallel to BC.

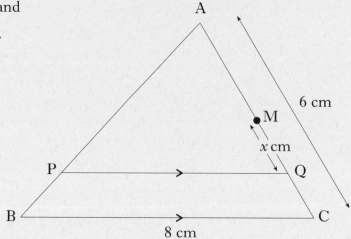

M is the midpoint of AC.

Q lies on AC, $x$ centimetres from M, as shown on the diagram.

(a) Write down an expression for the length of AQ.    1

(b) Show that PQ = $(4 + \frac{4}{3}x)$ centimetres.    3

[END OF QUESTION PAPER]

STANDARD GRADE | CREDIT

2007

[BLANK PAGE]

C

# 2500/405

NATIONAL
QUALIFICATIONS
2007

THURSDAY, 3 MAY
1.30 PM – 2.25 PM

MATHEMATICS
STANDARD GRADE
Credit Level
Paper 1
(Non-calculator)

1 **You may NOT use a calculator**.

2 Answer as many questions as you can.

3 Full credit will be given only where the solution contains appropriate working.

4 Square-ruled paper is provided.

SCOTTISH
QUALIFICATIONS
AUTHORITY

©

**FORMULAE LIST**

The roots of $ax^2 + bx + c = 0$ are $x = \dfrac{-b \pm \sqrt{(b^2 - 4ac)}}{2a}$

**Sine rule:** $\dfrac{a}{\sin A} = \dfrac{b}{\sin B} = \dfrac{c}{\sin C}$

**Cosine rule:** $a^2 = b^2 + c^2 - 2bc \cos A$ or $\cos A = \dfrac{b^2 + c^2 - a^2}{2bc}$

**Area of a triangle:**    Area $= \frac{1}{2}ab \sin C$

**Standard deviation:**    $s = \sqrt{\dfrac{\sum(x - \bar{x})^2}{n-1}} = \sqrt{\dfrac{\sum x^2 - (\sum x)^2/n}{n-1}}$, where $n$ is the sample size.

| | KU | RE |
|---|---|---|

1. Evaluate

$$6 \cdot 04 + 3 \cdot 72 \times 20.$$

**2**

2. Evaluate

$$3\tfrac{1}{6} \div 1\tfrac{2}{3}.$$

**2**

3. There are 400 people in a studio audience.

The probability that a person chosen at random from this audience is male is $\tfrac{5}{8}$.

How many males are in this audience?

**2**

4.
$$P = \frac{2(m-4)}{3}$$

Change the subject of the formula to $m$.

**3**

5. Remove brackets and simplify

$$(2x + 3)^2 - 3(x^2 - 6).$$

**3**

**[Turn over**

**6.** A taxi fare consists of a £2 "call-out" charge **plus** a fixed amount per kilometre.

The graph shows the fare, $f$ pounds for a journey of $d$ kilometres.

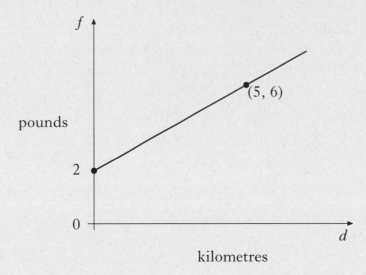

The taxi fare for a 5 kilometre journey is £6.

Find the equation of the straight line in terms of $d$ and $f$.

**7.** Remove brackets and simplify

$$a^{\frac{1}{2}}(a^{\frac{1}{2}} - 2).$$

KU RE

**8.** Mick needs an ironing board.

He sees one in a catalogue with measurements as shown in the diagram below.

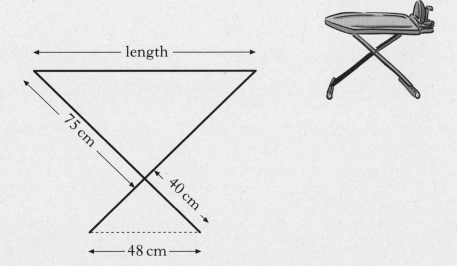

When the ironing board is set up, two similar triangles are formed.

Mick wants an ironing board which is at least 80 centimetres in length.

Does this ironing board meet Mick's requirements?

**Show all your working.**

3

**9.** A square of side $x$ centimetres has a diagonal 6 centimetres long.

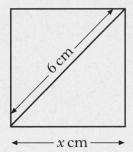

Calculate the value of $x$, giving your answer as a surd in its simplest form.

3

**[Turn over**

|  | KU | RE |
|---|---|---|

10. A relationship between $T$ and $L$ is given by the formula, $T = \dfrac{k}{L^3}$ where $k$ is a constant.

When $L$ is doubled, what is the effect on $T$?    **2**

11. (a) A cinema has 300 seats which are either standard or deluxe.

Let $x$ be the number of standard seats and $y$ be the number of deluxe seats.

Write down an algebraic expression to illustrate this information.    **1**

(b) A standard seat costs £4 and a deluxe seat costs £6.

When all the seats are sold the ticket sales are £1380.

Write down an algebraic expression to illustrate this information.    **2**

(c) How many standard seats and how many deluxe seats are in the cinema?    **3**

**12.** The diagram shows water lying in a length of roof guttering.

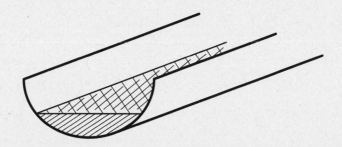

The cross-section of the guttering is a semi-circle with diameter 10 centimetres.

The water surface is 8 centimetres wide.

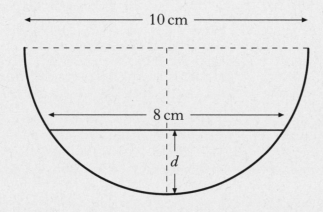

Calculate the depth, $d$, of water in the guttering.

4

**[Turn over for Questions 13 and 14 on *Page eight***

KU | RE

**13.** Part of the graph of $y = \cos bx° + c$ is shown below.

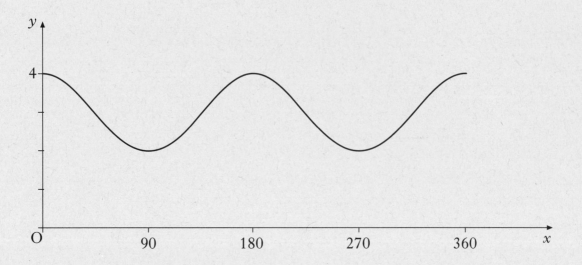

Write down the values of $b$ and $c$.

2

**14.** The **sum** $S_n$ of the first $n$ terms of a sequence, is given by the formula

$$S_n = 3^n - 1.$$

(a) Find the **sum** of the first 2 terms.

1

(b) When $S_n = 80$, calculate the value of $n$.

2

*[END OF QUESTION PAPER]*

**C**

# 2500/406

NATIONAL
QUALIFICATIONS
2007

THURSDAY, 3 MAY
2.45 PM – 4.05 PM

MATHEMATICS
STANDARD GRADE
Credit Level
Paper 2

1 **You may use a calculator**.

2 Answer as many questions as you can.

3 Full credit will be given only where the solution contains appropriate working.

4 Square-ruled paper is provided.

SCOTTISH
QUALIFICATIONS
AUTHORITY

©

**FORMULAE LIST**

The roots of $ax^2 + bx + c = 0$ are $x = \dfrac{-b \pm \sqrt{(b^2 - 4ac)}}{2a}$

**Sine rule:** $\dfrac{a}{\sin A} = \dfrac{b}{\sin B} = \dfrac{c}{\sin C}$

**Cosine rule:** $a^2 = b^2 + c^2 - 2bc \cos A$ or $\cos A = \dfrac{b^2 + c^2 - a^2}{2bc}$

**Area of a triangle:** Area $= \frac{1}{2}ab \sin C$

**Standard deviation:** $s = \sqrt{\dfrac{\sum(x - \bar{x})^2}{n-1}} = \sqrt{\dfrac{\sum x^2 - (\sum x)^2 / n}{n-1}}$, where $n$ is the sample size.

|  | KU | RE |
|---|---|---|

**1.** Alistair buys an antique chair for £600.

It is expected to increase in value at the rate of 4·5% each year.

How much is it expected to be worth in 3 years?    **3**

**2.** Solve the equation

$$3x^2 - 2x - 10 = 0.$$

Give your answer **correct to 2 significant figures**.    **4**

**3.** (a) During his lunch hour, Luke records the number of birds that visit his bird-table.

The numbers recorded last week were:

$$28 \quad 32 \quad 14 \quad 19 \quad 18 \quad 26 \quad 31.$$

Find the mean and standard deviation for this data.    **4**

(b) Over the same period, Luke's friend, Erin also recorded the number of birds visiting her bird-table.

Erin's recordings have a mean of 25 and a standard deviation of 5.

Make **two** valid comparisons between the friends' recordings.    **2**

**4.** Solve the inequality

$$\frac{x}{4} - \frac{1}{2} < 5.$$    **2**

**[Turn over**

KU | RE

5.  Mark takes some friends out for a meal.

The restaurant adds a 10% service charge to the price of the meal.

The **total** bill is £148·50.

What was the price of the meal?

3

6.  Brunton is 30 kilometres due North of Appleton.

From Appleton, the bearing of Carlton is 065°.

From Brunton, the bearing of Carlton is 153°.

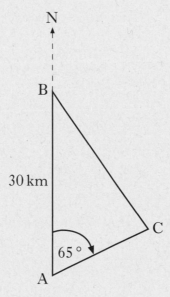

Calculate the distance between Brunton and Carlton.

4

KU | RE

**7.** A fan has four identical plastic blades.

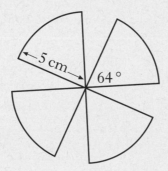

Each blade is a sector of a circle of radius 5 centimetres.

The angle at the centre of each sector is 64°.

Calculate the **total** area of plastic required to make the blades.

3

**8.** In triangle PQR:
- QR = 6 centimetres
- angle PQR = 30°
- area of triangle PQR = 15 square centimetres.

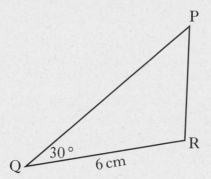

Calculate the length of PQ.

3

**[Turn over**

**9.** To make "14 carat" gold, copper and pure gold are mixed in the ratio 5:7.

A jeweller has 160 grams of copper and 245 grams of pure gold.

What is the maximum weight of "14 carat" gold that the jeweller can make?

**10.** Solve **algebraically** the equation

$$5 \cos x° + 4 = 0, \qquad 0 \le x < 360.$$

**11.** (*a*) A decorator's logo is rectangular and measures 10 centimetres by 6 centimetres.

It consists of three rectangles: one red, one yellow and one blue.

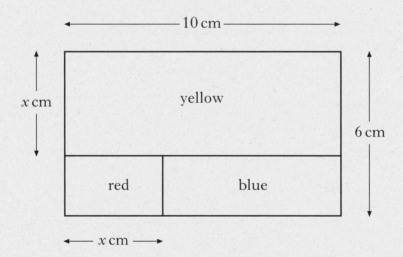

The yellow rectangle measures 10 centimetres by *x* centimetres.

The width of the red rectangle is *x* centimetres.

Show that the area, A, of the blue rectangle is given by the expression

$$A = x^2 - 16x + 60.$$

(*b*) The area of the blue rectangle is equal to $\frac{1}{5}$ of the total area of the logo.

Calculate the value of *x*.

KU | RE

**12.**    (*a*)   A cylindrical paperweight of radius 3 centimetres and height 4 centimetres is filled with sand.

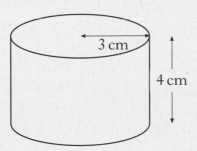

Calculate the volume of sand in the paperweight.

2

    (*b*)   Another paperweight, in the shape of a hemisphere, is filled with sand.

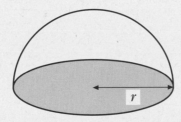

It contains the same volume of sand as the first paperweight.

Calculate the radius of the hemisphere.

[The volume of a hemisphere with radius *r* is given by the formula, $V = \frac{2}{3}\pi r^3$].

3

**[Turn over for Question 13 on *Page eight***

KU RE

**13.** The profit made by a publishing company of a magazine is calculated by the formula

$$y = 4x(140 - x),$$

where $y$ is the profit (in pounds) and $x$ is the selling price (in pence) of the magazine.

The graph below represents the profit $y$ against the selling price $x$.

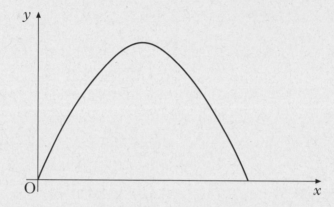

Find the maximum profit the company can make from the sale of the magazine.

4

*[END OF QUESTION PAPER]*

STANDARD GRADE | CREDIT

# 2008

[BLANK PAGE]

C

# 2500/405

NATIONAL
QUALIFICATIONS
2008

THURSDAY, 8 MAY
1.30 PM – 2.25 PM

MATHEMATICS
STANDARD GRADE
Credit Level
Paper 1
(Non-calculator)

1 **You may NOT use a calculator**.

2 Answer as many questions as you can.

3 Full credit will be given only where the solution contains appropriate working.

4 Square-ruled paper is provided.

**FORMULAE LIST**

The roots of $ax^2 + bx + c = 0$ are $x = \dfrac{-b \pm \sqrt{(b^2 - 4ac)}}{2a}$

**Sine rule:** $\dfrac{a}{\sin A} = \dfrac{b}{\sin B} = \dfrac{c}{\sin C}$

**Cosine rule:** $a^2 = b^2 + c^2 - 2bc \cos A$  or  $\cos A = \dfrac{b^2 + c^2 - a^2}{2bc}$

**Area of a triangle:**    Area $= \frac{1}{2} ab \sin C$

**Standard deviation:**    $s = \sqrt{\dfrac{\sum(x - \bar{x})^2}{n-1}} = \sqrt{\dfrac{\sum x^2 - (\sum x)^2 / n}{n-1}}$, where $n$ is the sample size.

KU | RE

1. Evaluate

$$24 \cdot 7 - 0 \cdot 63 \times 30.$$

2

2. Factorise fully

$$5x^2 - 45.$$

2

3. $$W = BH^2.$$

Change the subject of the formula to $H$.

2

4. A straight line cuts the $x$-axis at the point (9, 0) and the $y$-axis at the point (0, 18) as shown.

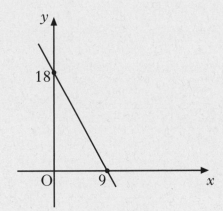

Find the equation of this line.

3

**[Turn over**

| | KU | RE |
|---|---|---|

**5.** Express as a single fraction in its simplest form

$$\frac{1}{p} + \frac{2}{(p+5)}.$$

KU **2**

**6.** Jane enters a two-part race.

(a) She cycles for 2 hours at a speed of $(x + 8)$ kilometres per hour.

Write down an expression in $x$ for the distance cycled.

KU **1**

(b) She then runs for 30 minutes at a speed of $x$ kilometres per hour.

Write down an expression in $x$ for the distance run.

KU **1**

(c) The **total** distance of the race is 46 kilometres.

Calculate Jane's **running** speed.

RE **3**

**7.** The 4th term of each number pattern below is the **mean** of the previous three terms.

(a) When the first three terms are 1, 6, and 8, calculate the 4th term.

KU **1**

(b) When the first three terms are $x$, $(x + 7)$ and $(x + 11)$, calculate the 4th term.

RE **1**

(c) When the first, second and fourth terms are

$$-2x, \qquad (x+5), \qquad \underline{\hspace{1cm}}, \qquad (2x+4),$$

calculate the 3rd term.

RE **2**

**8.** The curved part of the letter A in the *Artwork* logo is in the shape of a parabola.

The equation of this parabola is $y = (x - 8)(2 - x)$.

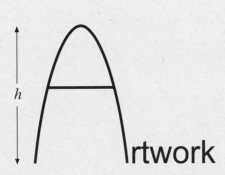

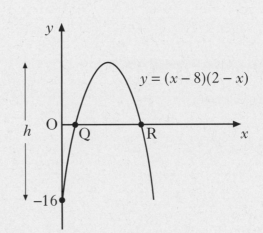

(*a*) Write down the coordinates of Q and R.

(*b*) Calculate the height, *h*, of the letter A.

**9.** Simplify

$$m^3 \times \sqrt{m}.$$

**[Turn over**

KU | RE

**10.** Part of the graph of $y = a^x$, where $a > 0$, is shown below.

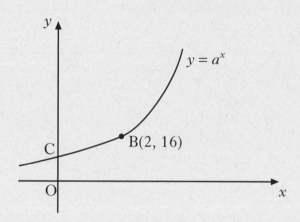

The graph cuts the $y$-axis at C.

(*a*) Write down the coordinates of C.

1

B is the point (2, 16).

(*b*) Calculate the value of $a$.

2

**11.** A right angled triangle has dimensions as shown.

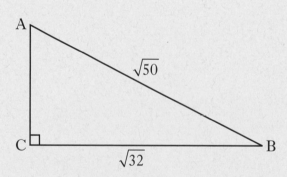

Calculate the length of AC, leaving your answer as a surd **in its simplest form**.

3

KU | RE

**12.** Given that

$$x^2 - 10x + 18 = (x - a)^2 + b,$$

find the values of $a$ and $b$.

3

**13.** A new fraction is obtained by adding $x$ to the numerator and denominator of the fraction $\frac{17}{24}$.

This new fraction is equivalent to $\frac{2}{3}$.

Calculate the value of $x$.

3

*[END OF QUESTION PAPER]*

[BLANK PAGE]

C

# 2500/406

NATIONAL
QUALIFICATIONS
2008

THURSDAY, 8 MAY
2.45 PM – 4.05 PM

MATHEMATICS
STANDARD GRADE
Credit Level
Paper 2

1  **You may use a calculator**.

2  Answer as many questions as you can.

3  Full credit will be given only where the solution contains appropriate working.

4  Square-ruled paper is provided.

# FORMULAE LIST

The roots of $ax^2 + bx + c = 0$ are $x = \dfrac{-b \pm \sqrt{(b^2 - 4ac)}}{2a}$

**Sine rule:** $\dfrac{a}{\sin A} = \dfrac{b}{\sin B} = \dfrac{c}{\sin C}$

**Cosine rule:** $a^2 = b^2 + c^2 - 2bc \cos A$ or $\cos A = \dfrac{b^2 + c^2 - a^2}{2bc}$

**Area of a triangle:**   Area $= \frac{1}{2}ab \sin C$

**Standard deviation:**   $s = \sqrt{\dfrac{\sum(x - \bar{x})^2}{n-1}} = \sqrt{\dfrac{\sum x^2 - (\sum x)^2 / n}{n-1}}$, where $n$ is the sample size.

KU | RE

1. A local council recycles 42 000 tonnes of waste a year.

   The council aims to increase the amount of waste recycled by 8% each year.

   How much waste does it expect to recycle in 3 years time?

   Give your answer **to three significant figures**.

4

2. In a class, 30 pupils sat a test.

   The marks are illustrated by the stem and leaf diagram below.

   **Test Marks**

   | 0 | 9 |
   |---|---|
   | 1 | 6  6  7  8 |
   | 2 | 0  4  5  7  9  9  9 |
   | 3 | 2  2  3  5  5  6  8 |
   | 4 | 0  2  3  4  5  5  7  7  8 |
   | 5 | 0  0 |

   n = 30                              1 | 6 = 16

   (a) Write down the median and the modal mark.

2

   (b) Find the probability that a pupil selected at random scored **at least** 40 marks.

1

3. In a sale, all cameras are reduced by 20%.

   A camera now costs £45.

   Calculate the **original** cost of the camera.

NOW
£45

3

**[Turn over**

**4.** Aaron saves 50 pence and 20 pence coins in his piggy bank.

Let $x$ be the number of 50 pence coins in his bank.

Let $y$ be the number of 20 pence coins in his bank.

(*a*) There are 60 coins in his bank.

Write down an equation in $x$ and $y$ to illustrate this information.

(*b*) The total value of the coins is £17·40.

Write down another equation in $x$ and $y$ to illustrate this information.

(*c*) Hence find **algebraically** the number of 50 pence coins Aaron has in his piggy bank.

**5.** A circle, centre the origin, is shown.

P is the point (8, 1).

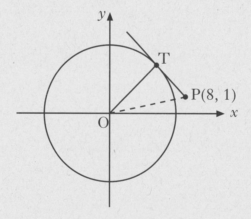

(*a*) Calculate the length of OP.

The diagram also shows a tangent from P which touches the circle at T.

The radius of the circle is 5 units.

(*b*) Calculate the length of PT.

| | KU | RE |
|---|---|---|

**6.** The distance, *d* kilometres, to the horizon, when viewed from a cliff top, varies directly as the square root of the height, *h* metres, of the cliff top above sea level.

From a cliff top 16 metres above sea level, the distance to the horizon is 14 kilometres.

A boat is 20 kilometres from a cliff whose top is 40 metres above sea level.

Is the boat beyond the horizon?

**Justify your answer.**

5

**7.** A telegraph pole is 6·2 metres high.

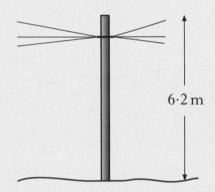

6·2 m

The wind blows the pole over into the position as shown below.

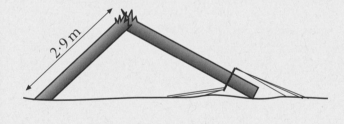

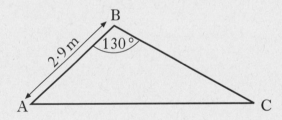

AB is 2·9 metres and angle ABC is 130°.

Calculate the length of AC.

4

**[Turn over**

KU | RE

**8.** A farmer builds a sheep-pen using two lengths of fencing and a wall.

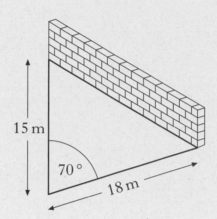

15 m

70°

18 m

The two lengths of fencing are 15 metres and 18 metres long.

(a) Calculate the area of the sheep-pen, when the angle between the fencing is 70°.

3

(b) What angle between the fencing would give the farmer the largest possible area?

1

**9.** Contestants in a quiz have 25 seconds to answer a question.

This time is indicated on the clock.

The tip of the clock hand moves through the arc AB as shown.

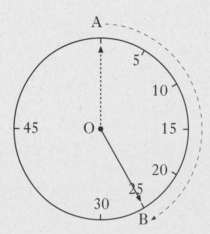

(a) Calculate the size of angle AOB.

1

(b) The length of arc AB is 120 centimetres.

Calculate the length of the clock hand.

4

KU | RE

**10.** To hire a car costs £25 per day plus a mileage charge.

The first 200 miles are free with each additional mile charged at 12 pence.

> # CAR HIRE
>
> ## £25 per day
>
> - **first 200** miles free
> - each additional mile only 12p

(a) Calculate the cost of hiring a car for 4 days when the mileage is 640 miles.

1

(b) A car is hired for $d$ days and the mileage is $m$ miles where $m > 200$.

Write down a formula for the cost £$C$ of hiring the car.

3

**11.** The minimum number of roads joining 4 towns to each other is 6 as shown.

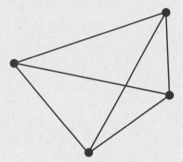

The minimum number of roads, $r$, joining $n$ towns to each other is given by the formula
$$r = \tfrac{1}{2}n(n-1).$$

(a) State the minimum number of roads needed to join 7 towns to each other.

1

(b) When $r = 55$, show that $n^2 - n - 110 = 0$.

2

(c) Hence find **algebraically** the value of $n$.

3

**[Turn over for Question 12 on *Page eight***

KU | RE

**12.** The diagram shows part of the graph of $y = \tan x°$.

The line $y = 5$ is drawn and intersects the graph of $y = \tan x°$ at P and Q.

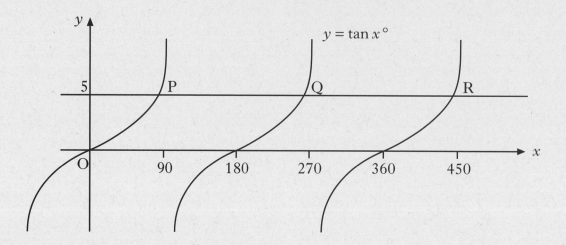

(a) Find the $x$-coordinates of P and Q.

3

(b) Write down the $x$-coordinate of the point R, where the line $y = 5$ next intersects the graph of $y = \tan x°$.

1

*[END OF QUESTION PAPER]*

# STANDARD GRADE | CREDIT

# 2009

[BLANK PAGE]

C

# 2500/405

NATIONAL
QUALIFICATIONS
2009

WEDNESDAY, 6 MAY
1.30 PM – 2.25 PM

MATHEMATICS
STANDARD GRADE
Credit Level
Paper 1
(Non-calculator)

1 **You may NOT use a calculator**.

2 Answer as many questions as you can.

3 Full credit will be given only where the solution contains appropriate working.

4 Square-ruled paper is provided.

**FORMULAE LIST**

The roots of $ax^2 + bx + c = 0$ are $x = \dfrac{-b \pm \sqrt{(b^2 - 4ac)}}{2a}$

**Sine rule:** $\dfrac{a}{\sin A} = \dfrac{b}{\sin B} = \dfrac{c}{\sin C}$

**Cosine rule:** $a^2 = b^2 + c^2 - 2bc \cos A$ or $\cos A = \dfrac{b^2 + c^2 - a^2}{2bc}$

**Area of a triangle:** Area $= \frac{1}{2}ab \sin C$

**Standard deviation:** $s = \sqrt{\dfrac{\sum(x - \bar{x})^2}{n - 1}} = \sqrt{\dfrac{\sum x^2 - (\sum x)^2 / n}{n - 1}}$, where $n$ is the sample size.

KU | RE

**1.** Evaluate

$$(846 \div 30) - 1 \cdot 09.$$

2

**2.** Evaluate

$$4\frac{1}{3} - 1\frac{1}{2} .$$

2

**3.** Given that

$$f(x) = x^2 + 3,$$

(a) evaluate $f(-4)$

2

(b) find $t$ when $f(t) = 52$.

2

**4.** (a) Factorise

$$x^2 - 4y^2.$$

1

(b) Expand and simplify

$$(2x - 1)(x + 4).$$

1

(c) Expand

$$x^{\frac{1}{2}}\left(3x + x^{-2}\right).$$

2

**[Turn over**

KU | RE

**5.** In triangle ABC:

- angle ACB = 90°

- AB = 8 centimetres

- AC = 4 centimetres.

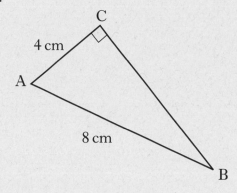

Calculate the length of BC.

Give your answer **as a surd in its simplest form**.

3

**6.** There are 4 girls and 14 boys in a class.

A child is chosen at random and is asked to roll a die, numbered 1 to 6.

Which of these is more likely?

    A: the child is female.

      **OR**

    B: the child rolls a 5.

**Justify your answer.**

3

**7.** This year, Ben paid £260 for his car insurance.

This is an increase of 30% on last year's payment.

How much did Ben pay last year?

3

KU | RE

**8.** In triangle PQR:

- PQ = $x$ centimetres
- PR = $5x$ centimetres
- QR = $2y$ centimetres.

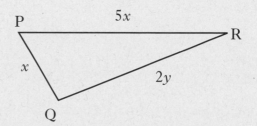

(a) The perimeter of the triangle is 42 centimetres.

Write down an equation in $x$ and $y$ to illustrate this information.

2

(b) PR is 2 centimetres longer than QR.

Write down another equation in $x$ and $y$ to illustrate this information.

2

(c) Hence calculate the values of $x$ and $y$.

3

**9.** A formula used to calculate the flow of water in a pipe is

$$f = \frac{kd^2}{20}.$$

Change the subject of the formula to $d$.

3

**[Turn over**

**10.** The diagram below shows the path of a rocket which is fired into the air.

The height, $h$ metres, of the rocket after $t$ seconds is given by

$$h(t) = -2t(t - 14).$$

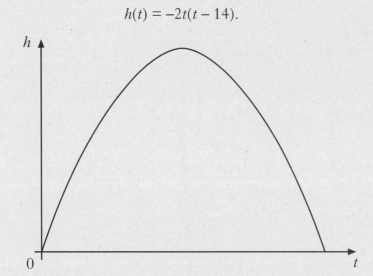

(a) For how many seconds is the rocket in flight?    2

(b) What is the maximum height reached by the rocket?    2

KU | RE

**11.** In triangle ABC:

- BC = 6 metres

- AC = 10 metres

- angle ABC = 30°.

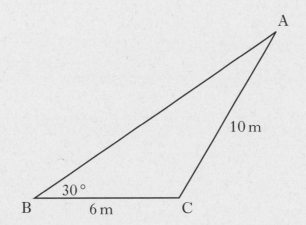

Given that sin 30° = 0·5, show that sin A = 0·3.

3

*[END OF QUESTION PAPER]*

KU | RE

[BLANK PAGE]

C

# 2500/406

NATIONAL
QUALIFICATIONS
2009

WEDNESDAY, 6 MAY
2.45 PM – 4.05 PM

MATHEMATICS
STANDARD GRADE
Credit Level
Paper 2

1 **You may use a calculator**.

2 Answer as many questions as you can.

3 Full credit will be given only where the solution contains appropriate working.

4 Square-ruled paper is provided.

**FORMULAE LIST**

The roots of $ax^2 + bx + c = 0$ are $x = \dfrac{-b \pm \sqrt{(b^2 - 4ac)}}{2a}$

**Sine rule:** $\dfrac{a}{\sin A} = \dfrac{b}{\sin B} = \dfrac{c}{\sin C}$

**Cosine rule:** $a^2 = b^2 + c^2 - 2bc \cos A$   or   $\cos A = \dfrac{b^2 + c^2 - a^2}{2bc}$

**Area of a triangle:**    Area $= \frac{1}{2} ab \sin C$

**Standard deviation:**    $s = \sqrt{\dfrac{\sum (x - \bar{x})^2}{n - 1}} = \sqrt{\dfrac{\sum x^2 - (\sum x)^2 / n}{n - 1}}$, where $n$ is the sample size.

KU | RE

1. One atom of gold weighs $3 \cdot 27 \times 10^{-22}$ grams.

   How many atoms will there be in one kilogram of gold?

   Give your answer **in scientific notation correct to 2 significant figures**.

   3

2. Lemonade is to be poured from a 2 litre bottle into glasses.

   Each glass is in the shape of a cylinder of radius 3 centimetres and height 8 centimetres.

   How many full glasses can be poured from the bottle?

   4

3. Solve the quadratic equation $x^2 - 4x - 6 = 0$.

   Give your answers **correct to 1 decimal place**.

   4

**[Turn over**

KU RE

**4.** Two fridge magnets are mathematically similar.

One magnet is 4 centimetres long and the other is 10 centimetres long.

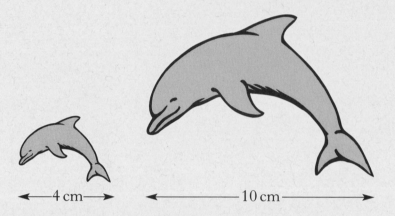

←—4 cm—→          ←————— 10 cm —————→

The area of the smaller magnet is 18 square centimetres.

Calculate the area of the larger magnet.

3

**5.** Tom looked at the cost of 10 different flights to New York.

He calculated that the mean cost was £360 and the standard deviation was £74.

A tax of £12 is then added to each flight.

Write down the new mean and standard deviation.

2

KU | RE

**6.** Teams in a quiz answer questions on film and sport.

This scatter graph shows the scores of some of the teams.

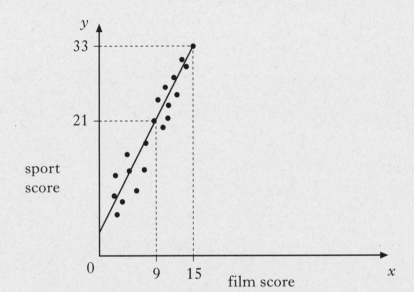

A line of best fit is drawn as shown above.

(*a*) Find the equation of this straight line.

4

(*b*) Use this equation to estimate the sport score for a team with a film score of 20.

2

**7.** (*a*) The air temperature, $t°$ Celsius, varies inversely as the square of the distance, $d$ metres, from a furnace.

Write down a formula connecting $t$ and $d$.

2

(*b*) At a distance of 2 metres from the furnace, the air temperature is 50 °C.

Calculate the air temperature at a distance of 5 metres from the furnace.

3

**[Turn over**

**8.** A company makes large bags of crisps which contain 90 grams of fat.

The company aims to reduce the fat content of the crisps by 50%.

They decide to reduce the fat content by 20% each year.

Will they have achieved their aim by the end of the 3rd year?

**Justify your answer.**

4

**9.** Jane is taking part in an orienteering competition.

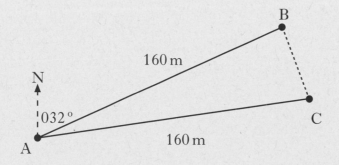

She should have run 160 metres from A to B on a bearing of 032°.

However, she actually ran 160 metres from A to C on a bearing of 052°.

(*a*) Write down the size of angle BAC.

1

(*b*) Calculate the length of BC.

3

(*c*) What is the bearing from C to B?

2

KU | RE

**10.** The weight, $W$ kilograms, of a giraffe is related to its age, $M$ months, by the formula

$$W = \tfrac{1}{4}\left(M^2 - 4M + 272\right).$$

At what age will a giraffe weigh 83 kilograms?

4

**11.** A cone is formed from a paper circle with a sector removed as shown.

The radius of the paper circle is 30 cm.

Angle AOB is 100°.

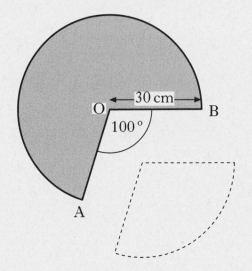

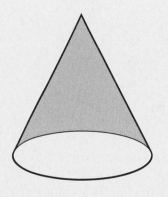

(a) Calculate the area of paper used to make the cone.

3

(b) Calculate the circumference of the base of the cone.

3

**[Turn over for Question 12 on *Page eight***

**12.** The $n^{\text{th}}$ term, $T_n$ of the sequence 1, 3, 6, 10, . . . is given by the formula:

$$T_n = \tfrac{1}{2}n(n+1)$$

1st term $\quad T_1 = \tfrac{1}{2} \times 1(1+1) = 1$

2nd term $\quad T_2 = \tfrac{1}{2} \times 2(2+1) = 3$

3rd term $\quad T_3 = \tfrac{1}{2} \times 3(3+1) = 6$

(a) Calculate the $20^{\text{th}}$ term, $T_{20}$.

1

(b) Show that $\quad T_{n+1} = \tfrac{1}{2}\left(n^2 + 3n + 2\right)$.

2

(c) Show that $T_n + T_{n+1}$ is a square number.

2

*[END OF QUESTION PAPER]*

[BLANK PAGE]

C

# 2500/405

NATIONAL
QUALIFICATIONS
2010

WEDNESDAY, 5 MAY
1.30 PM – 2.25 PM

MATHEMATICS
STANDARD GRADE
Credit Level
Paper 1
(Non-calculator)

1 **You may NOT use a calculator**.

2 Answer as many questions as you can.

3 Full credit will be given only where the solution contains appropriate working.

4 Square-ruled paper is provided.

**FORMULAE LIST**

The roots of $ax^2 + bx + c = 0$ are $x = \dfrac{-b \pm \sqrt{(b^2 - 4ac)}}{2a}$

**Sine rule:** $\dfrac{a}{\sin A} = \dfrac{b}{\sin B} = \dfrac{c}{\sin C}$

**Cosine rule:** $a^2 = b^2 + c^2 - 2bc \cos A$   or   $\cos A = \dfrac{b^2 + c^2 - a^2}{2bc}$

**Area of a triangle:**    Area $= \frac{1}{2} ab \sin C$

**Standard deviation:**    $s = \sqrt{\dfrac{\sum(x - \bar{x})^2}{n-1}} = \sqrt{\dfrac{\sum x^2 - (\sum x)^2 / n}{n-1}}$, where $n$ is the sample size.

| | KU | RE |
|---|---|---|

**1.** Evaluate

$$40\% \text{ of } £11{\cdot}50 - £1{\cdot}81.$$

2

**2.** Evaluate

$$\frac{2}{5} \div 1\frac{1}{10}.$$

2

**3.** Change the subject of the formula to $s$.

$$t = \frac{7s + 4}{2}.$$

3

**4.** Two functions are given below.

$$f(x) = x^2 - 4x$$

$$g(x) = 2x + 7$$

(a) If $f(x) = g(x)$, show that $x^2 - 6x - 7 = 0$.

2

(b) Hence find **algebraically** the values of $x$ for which $f(x) = g(x)$.

2

**[Turn over**

KU | RE

**5.** A bag contains 27 marbles.  Some are black and some are white.

The probability that a marble chosen at random is black is $\frac{4}{9}$.

(a) What is the probability that a marble chosen at random is white?

(b) How many white marbles are in the bag?

**6.** Cleano washing powder is on special offer.

Each box on special offer contains 20% more powder than the standard box.

A box on special offer contains 900 grams of powder.

How many grams of powder does the standard box contain?

| | KU | RE |
|---|---|---|

7. A straight line has equation $y = mx + c$, where $m$ and $c$ are constants.

   (a) The point (2, 7) lies on this line.

   Write down an equation in $m$ and $c$ to illustrate this information. **1**

   (b) A second point (4, 17) also lies on this line.

   Write down another equation in $m$ and $c$ to illustrate this information. **1**

   (c) Hence calculate the values of $m$ and $c$. **3**

   (d) Write down the gradient of this line. **1**

8. (a) Simplify $\sqrt{2} \times \sqrt{18}$. **1**

   (b) Simplify $\sqrt{2} + \sqrt{18}$. **1**

   (c) Hence show that $\dfrac{\sqrt{2} \times \sqrt{18}}{\sqrt{2} + \sqrt{18}} = \dfrac{3\sqrt{2}}{4}$ . **2**

**[Turn over**

KU | RE

**9.** Part of the graph of the straight line with equation $y = \frac{1}{3}x + 2$, is shown below.

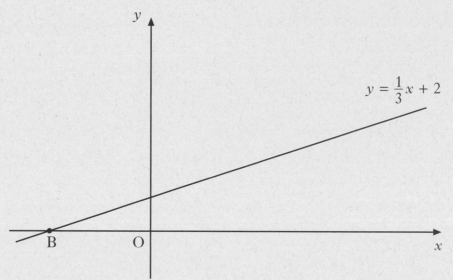

(a) Find the coordinates of the point B.

2

(b) For what values of $x$ is $y < 0$?

1

KU | RE

**10.** A number pattern is shown below.

$$1^3 = \frac{1^2 \times 2^2}{4}$$

$$1^3 + 2^3 = \frac{2^2 \times 3^2}{4}$$

$$1^3 + 2^3 + 3^3 = \frac{3^2 \times 4^2}{4}$$

(*a*) Write down a similar expression for $1^3 + 2^3 + 3^3 + 4^3 + 5^3$.

1

(*b*) Write down a similar expression for $1^3 + 2^3 + 3^3 + \ldots + n^3$.

2

(*c*) Hence **evaluate** $1^3 + 2^3 + 3^3 + \ldots + 9^3$.

2

**11.** Two triangles have dimensions as shown.

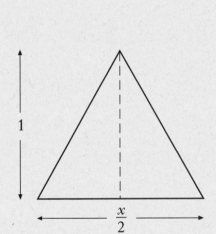
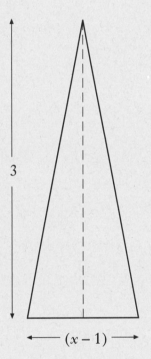

The triangles are equal in area.

**Calculate** the value of $x$.

4

*[END OF QUESTION PAPER]*

[BLANK PAGE]

C

## 2500/406

| | | |
|---|---|---|
| NATIONAL QUALIFICATIONS 2010 | WEDNESDAY, 5 MAY 2.45 PM – 4.05 PM | **MATHEMATICS** STANDARD GRADE Credit Level Paper 2 |

1 **You may use a calculator**.

2 Answer as many questions as you can.

3 Full credit will be given only where the solution contains appropriate working.

4 Square-ruled paper is provided.

**FORMULAE LIST**

The roots of $ax^2 + bx + c = 0$ are $x = \dfrac{-b \pm \sqrt{(b^2 - 4ac)}}{2a}$

**Sine rule:** $\dfrac{a}{\sin A} = \dfrac{b}{\sin B} = \dfrac{c}{\sin C}$

**Cosine rule:** $a^2 = b^2 + c^2 - 2bc \cos A$ or $\cos A = \dfrac{b^2 + c^2 - a^2}{2bc}$

**Area of a triangle:** Area $= \frac{1}{2}ab \sin C$

**Standard deviation:** $s = \sqrt{\dfrac{\sum(x - \bar{x})^2}{n-1}} = \sqrt{\dfrac{\sum x^2 - (\sum x)^2 / n}{n-1}}$, where $n$ is the sample size.

KU | RE

1. It is estimated that an iceberg weighs 84 000 tonnes.

   As the iceberg moves into warmer water, its weight decreases by 25% each day.

   What will the iceberg weigh after 3 days in the warmer water?

   Give your answer **correct to three significant figures**.

   4

2. Expand fully and simplify

   $$x(x-1)^2.$$

   2

3. A machine is used to put drawing pins into boxes.

   A sample of 8 boxes is taken and the number of drawing pins in each is counted.

   The results are shown below:

   | 102 | 102 | 101 | 98 | 99 | 101 | 103 | 102 |

   (a) Calculate the mean and standard deviation of this sample.

   3

   (b) A sample of 8 boxes is taken from another machine.

   This sample has a mean of 103 and a standard deviation of 2·1.

   Write down two valid comparisons between the samples.

   2

4. Use the quadratic formula to solve the equation,

   $$3x^2 + 5x - 7 = 0.$$

   Give your answers correct to **1 decimal place**.

   4

   **[Turn over**

**5.** A concrete ramp is to be built.

The ramp is in the shape of a cuboid and a triangular prism with dimensions as shown.

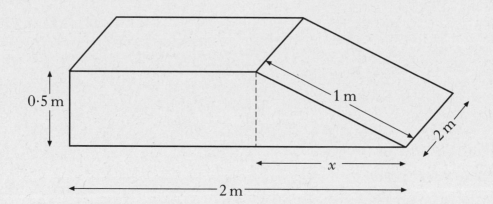

(*a*) Calculate the value of *x*.

(*b*) Calculate the volume of concrete required to build the ramp.

**6.** A circle, centre O, has radius 36 centimetres.

Part of this circle is shown.

Angle AOB = 140°.

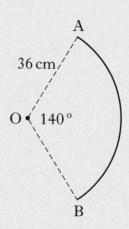

Calculate the length of arc AB.

7. Shampoo is available in travel size and salon size bottles.

   The bottles are mathematically similar.

travel           salon

The travel size contains 200 millilitres and is 12 centimetres in height.

The salon size contains 1600 millilitres.

Calculate the height of the salon size bottle.

3

[Turn over

**8.** As part of their training, footballers run around a triangular circuit DEF.

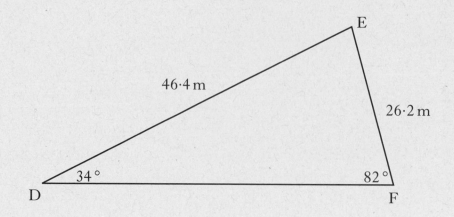

- ∠EDF = 34°

- ∠DFE = 82°

- DE = 46·4 metres

- EF = 26·2 metres

How many **complete** circuits must they run to cover **at least** 1000 metres?

**9.** The ratio of sugar to fruit in a particular jam is 5 : 4.

It is decided to:

- **decrease** the sugar content by 20%

- **increase** the fruit content by 20%.

Calculate the new ratio of sugar to fruit.

**Give your answer in its simplest form.**

KU | RE

**10.** In triangle PQR:

- PQ = 5 centimetres

- PR = 6 centimetres

- area of triangle PQR = 12 square centimetres

- angle QPR is **obtuse**.

Calculate the size of angle QPR.

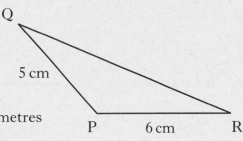

4

**11.** The height, $h$, of a square-based pyramid varies directly as its volume, $V$, and inversely as the square of the length of the base, $b$.

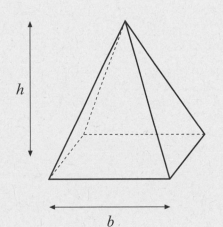

(a) Write down an equation connecting $h$, $V$ and $b$.

2

A square-based pyramid of height 12 centimetres has a volume of 256 cubic centimetres and length of base 8 centimetres.

(b) Calculate the height of a square-based pyramid of volume 600 cubic centimetres and length of base 10 centimetres.

3

**[Turn over for Questions 12 and 13 on *Page eight***

**12.** A right-angled triangle has dimensions, in centimetres, as shown.

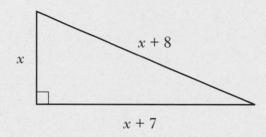

**Calculate** the value of $x$.

5

**13.** The depth of water, $D$ metres, in a harbour is given by the formula

$$D = 3 + 1 \cdot 75 \sin 30 h^\circ$$

where $h$ is the number of hours after midnight.

(a) Calculate the depth of water at 5 am.

2

(b) Calculate the maximum difference in depth of the water in the harbour.

**Do not use a trial and improvement method.**

2

*[END OF QUESTION PAPER]*

[BLANK PAGE]

[BLANK PAGE]

[BLANK PAGE]

[BLANK PAGE]